CBD Oil

M000240282

Relief

Ultimate Guide On Hemp Oil To Help You Relief Anxiety, Acne, Nausea, Obesity, Sleep Disorders, Stress, Stroke, Arthritis, Inflammation, Cardiovascular Diseases.

Gabriella Brown

Table of Contents

Copyright © 2021 Gabriella Brown

All rights reserved. No part of this publication may be reproduced, distributed, or transmitted in any form or by any means, including photocopying, recording, or other electronic or mechanical methods, without the prior written permission of the publisher, except in the case of brief quotations embodied in critical reviews and specific other non-commercial uses permitted by copyright law.

INTRODUCTION

People who have used CBD Oil traditionally used it for a large number of years to take care of numerous kinds of pains, however the medical community have only recently begun to review it again.

One report discovered that short-term use of CBD essential oil could decrease the degrees of spasticity a person feels.

For many individuals experiencing chronic pain, cannabidiol (CBD) oil has steadily gained recognition as an all-natural approach to treatment. A compound within the marijuana herb, cannabidiol may also be touted instead of pain medication in treating common conditions like arthritis and back pain.

CBD Oil has been talked about relentlessly in the media, online and all over society in recent months. Touted for being a *miracle drug*, CBD has worked its way into food, creams and even facials, making it the *most exiting supplement of 2018.*

Discover all the ways in which cannabidiol, a natural remedy from the cannabis plant, can change your life. The healing properties of this ancient medicine can boost mood, relieve pain, calm inflammation, improve heart health, strengthen bones, promote brain health, balance hormones, regulate the immune system, soothe skin conditions, and contribute to overall wellness in so many ways.

CHAPTER 1

Why People Use CBD Oil

Based on the Institute of Medication from the Country wide Academies, 100 million US citizens live with chronic pain. Along with significantly reducing standard of living, chronic pain can increase health care costs and also have a negative effect on productivity at the job.

Common types of chronic pain include:

- Cancer pain.

- Fibromyalgia.

- Headaches.

- Irritable bowel syndrome (IBS).

- Low back again pain.

- Migraines.

- Multiple sclerosis pain.

- Neuropathic pain.

- Osteoarthritis.

- Temporomandibular disorder (also known as "TMJ")

Over-the-counter and prescription pain medications tend to be recommended in the treating chronic pain, but many people look for alternative types of alleviation (such as natural herbs, natural supplements, and products like CBD essential oil).

A few of these people desire to avoid the medial side results frequently associated with standard pain medication, while some have concerns about becoming reliant on such medications. Actually, some proponents claim that CBD essential oil could give a treatment for opioid dependency as concerns over opioid overdoses continue steadily to escalate.

Potential Great things about CBD Oil

Scientists remain trying to regulate how CBD essential oil might alleviate pain. However, there's some proof that cannabidiol may impact the body's endocannabinoid system (a complicated system of cell-to-cell communication). Along with adding to brain functions like memory space and feeling, the endocannabinoid system affects how exactly we experience pain.

So far, a lot of the data for CBD's effects on pain management originates from animal-based research. When used orally, CBD has poor bioavailability. Topical CBD software to localized regions of pain is thought to provide more constant degrees of CBD with less systemic participation.

This research carries a study published in the journal Pain 2017, where scientists observed that treatment with topical CBD helped thwart the introduction of joint pain rats with osteoarthritis.

Another study, posted in the Western Journal of Pain 2016, discovered that topical CBD gel significantly reduced joint swelling and steps of pain and inflammation in rats with arthritis.

In a written report published in Pediatric Dermatology in 2018, scientists reported three cases of topical CBD (applied as an oil, cream, and apply) use in children with an uncommon, blistering condition of the skin known as epidermolysis bullosa. Applied by their parents, all three people reported faster wound curing, less blisters, and improvement of pain. One individual could completely wean off dental opioid analgesic pain medication. There have been no undesirable effects reported.

While hardly any clinical tests have explored the pain-relieving ramifications of CBD essential oil, a written report published in the Cochrane Database of Systematic Reviews in 2018 examined the utilization of a number of cannabis-based medications and found they could be of some benefit in the treading of chronic neuropathic pain. A kind of pain brought on by harm to the somatosensory system (i.e., the machine responsible for controlling sensory stimuli), neuropathic pain often occurs in people who have conditions like diabetes and multiple sclerosis.

In this record, researchers examined 16 previously published studies testing the utilization of varied cannabis-based medications in the treating chronic neuropathic pain and found some evidence that cannabis-based medications can help with treatment and decrease pain strength, rest difficulties, and psychological distress. Unwanted effects included sleepiness, dizziness, mental confusion. The authors figured the potential damage of such medications may outweigh their possible advantage, however, it ought to be mentioned that the studies used a number of cannabis-based medications (e.g. inhaled cannabis and sprays and dental tablets made up of THC and/or CBD from flower resources or made synthetically), a few of which will lead to these side results than products without THC.

Unwanted Effects and Safety

The research privately ramifications of CBD oil is incredibly limited. Furthermore, because of the lack of rules of several products, there is certainly inconsistency in this content and purity. The quantity of CBD in products might not be constant and products can contain differing levels of the psychoactive component THC.

While CBD is definitely the major non-psychoactive element of cannabis, in studies using varied dosages, routes of administration, and mixture or whole products with THC, lots of side results have been reported, including anxiety, changes in hunger and disposition, diarrhea, dizziness, drowsiness, dry mouth area, low blood circulation pressure, mental confusion, nausea, and vomiting.

There's also some concern that taking high dosages of cannabidiol could make muscle motion and tremors worse in people who have Parkinson's disease.

What's more, CBD essential oil may connect to certain medications, such as medications transformed by the liver organ (including chlorzoxazone, theophylline, clozapine, and progesterone) and sedative medications (including benzodiazepines, phenobarbital, fentanyl, and morphine).

When smoked, cannabis has been found to contain Aspergillus (a kind of fungus). People who have suppressed immune system systems should become aware of the chance of fungal contamination when working with this form of cannabis.

Topical CBD application could cause skin irritation.

CBD oil shouldn't be used as an alternative for standard treatment. Regarding chronic inflammatory conditions like arthritis, for example, chronic inflammation can result in joint harm (causing damage and impairment) if the problem is not effectively handled.

The Option of CBD Oil

As increasingly more states over the U.S. legalize the utilization of weed, CBD oil is becoming more accessible. CBD oil is currently bought from a variety of forms, including pills, creams, tinctures, and under-the-tongue sprays.

Even though many companies now sell CBD oil online

and in dispensaries, use of the oil isn't level atlanta divorce attorneys state. Because condition laws and regulations vary greatly as it pertains to cannabis products, it's essential to concur that use of CBD essential oil is legal in a state.

chapter 2

How Cannabis Essential oil Works

Relating to Medical Information Today, the body's endocannabinoid system has two cannabinoid receptors: CB1 and CB2.

Most CB1 receptors are situated in the mind and are associated with cognitive actions related to coordination, feeling, thinking, storage, and appetite. The CB2 receptors, on the other hands, are available in the disease fighting capability. This makes them more accountable for the body's response to pain and irritation.

Tetrahydrocannabinol, or THC, attaches to the CB1 receptors, which is why smoking or elsewhere ingesting marijuana impacts users on the cognitive level.

However, CBD effects the CB2 receipt. And it can so indirectly, not by attaching to the CB2 receptor, but by tempting your body to make more of its cannabinoids.

This creates a positive influence on the body's pain and swelling responses.

Great things about CBD Oil

Many reports have been conducted on CBD and discovered that this chemical substance offers individuals identified as having arthritis several potential benefits.

CBD Essential oil Eases Arthritis Pain

Among the main CBD essential oil benefits for arthritis victims, it is positive influence on pain, and research confirms it.

A report published in the journal Pain Dec 2017 analyzed whether CBD could prevent osteoarthritis pain and joint neuropathy. Predicated on their results, researchers affirmed it does both since it decreased joint irritation and offered as a protectant to the nerves.

CBD Essential oil Relieves Other Chronic Pain Conditions

CBD essential oil has been found to alleviate other

chronic aches and pains as well. For example, research released in Therapeutics and Clinical Risk Management shows that cannabinoids have been helpful with easing pain for folks identified as having conditions such as multiple sclerosis and malignancy.

Other studies have reported results for folks taking CBD to help ease their fibromyalgia pain, a lot of whom only experienced moderate side effects out of this chemical substance, such as dried out mouth area, drowsiness, and dizziness.

CBD Essential oil and Anxiety

CBD essential oil in addition has been associated with a number of mental health advantages, like a reduction in stress. Research shows that it is so effective in this respect it has "substantial potential as cure for multiple panic disorders." One research released in The Permanente Journal even discovered that CBD essential oil can lessen anxiousness in small children.

CBD Essential oil for Depression

Several studies also have connected CBD to a reduced number of symptoms associated with depression. That is vitally important because as the Arthritis Basis says that the depression rates of these identified as having certain types of arthritis "can be between two- and ten-times higher than the rates of the overall population."

Is CBD Essential oil Legal?

Although CBD oil has lots of potential benefits, one of the very best questions that folks diagnosed with arthritis rheumatoid have is if it is legal. Answering this question requires some knowledge of certain different parts of the marijuana vegetable.

The compound in the cannabis plant that is most well-known is THC. This is actually the chemical accountable for marijuana's notorious high. However, unlike THC, CBD doesn't have psychoactive properties. Which means that it generally does not create the same impact that one normally gets when smoking, inhaling, or elsewhere ingesting cannabis. Furthermore, the U.S. Medication Enforcement Administration (DEA) says that it's the THC which makes weed illegal in several different states.

Because of this, the response to "Is CBD oil legal?" isn't quite so clear.

Originally, marijuana use and sales became unlawful under the Marijuana Tax Take action of 1937, with this illegality continuing under the Controlled Chemicals Act. Yet, since that time, several states took the steps to legalize this medication, even if simply for use as medical cannabis.

For example, some says allow the usage of marijuana and its own extracts limited to medical reasons, making a medical marijuana cards essential for its use. Some claims have approved this medication and its own components for recreational use as well.

To make issues more technical, on Dec 14, 2016, the DEA issued your final Guideline that created a fresh code quantity (7350) made to cover "any materials, compound, combination, or preparation" surrounding the cannabis plant.

Upon clarification of the new code, the DEA indicated that CBD extracts which has even smaller amounts of other cannabinoids would still are categorized as the old code, making them unlawful under federal guideline. However, CBD components which contain no other cannabinoids are categorized as the new code, where they may potentially be legal to use.

Plus, some CBD natural oils derived from hemp, the fiber of the cannabis seed. Although 2014 Farm Expenses announced certain hemp cultivation methods legal, muddying the legal waters even more, the chief executive authorized the new Plantation Bill into legislation in Dec 2018, effectively legalizing hemp under federal government law.

What does all this mean in real-life conditions?

In a nutshell, to determine whether a CBD oil or any other CBD product is legal or unlawful depends on a variety of factors. Your physical location, local laws and regulations, and whether there are some other cannabinoids in the draw out must all be studied under consideration before coming to your final answer.

Is CBD Essential oil Safe?

The next most common question folks have as it pertains to CBD oil for pain management or its anti-inflammatory properties is if it is safe to use for arthritis symptoms.

Patients are always worried about additional potential and part effects (as they must be). Regarding for some top health experts and companies, the answer is yes. It really is safe to use.

For example, Medscape stocks that while cannabinoid medicine continues to be in its first stages, "unlike marijuana and THC, the potential risks associated with CBD are really low, with not really a single case report of CBD overdose in the literature. The National Institute on Substance Abuse agrees and states that "CBD is apparently a safe drug without addictive effects."

In the foreseeable future, chances are that the meals and Drug Administration (FDA) will also start to modify CBD products, providing yet another back-up, as this agency indicates that "increasing public interest" in it increases the need for establishing regulatory procedures.

Potential Risks of CBD Oil

Many health experts have deemed CBD safe to use, but as with any substance, there does look like a few potential risks as it pertains to by using this extract.

Medication Interactions

MedlinePlus indicates that CBD will often connect to certain prescription drugs, primarily the ones that are "changed and divided by the liver organ."

Included in these are, but aren't limited by common

medications such as:

- ondansetron (Zofran).

- clozapine (Clozaril, FazaClo).

- progesterone (Endometrin, Prometrium).

- Testosterone.

- disulfiram (Antabuse).

- ketamine (Ketalar).

- phenobarbital (Luminal).

- omeprazole (Prilosec).

- lansoprazole (Prevacid).

- diazepam (Valium).

- carisoprodol (Soma).

- ibuprofen (Motrin).

- celecoxib (Celebrex).

- amitriptyline (Elavil).

- warfarin (Coumadin).

- Codeine.

- paroxetine (Paxil).

- tramadol (Ultram).

- venlafaxine (Effexor).

- alprazolam (Xanax).

- fexofenadine (Allegra).

- Benzodiazepines.

- morphine

MedlinePlus further areas that we now have other "drugs" that could connect to CBD as well, including caffeine and nicotine. However, research hasn't necessarily backed these claims.

For example, a 2015 research posted in the journal Pharmacology Biochemistry and Behavior discovered that, in the right dose, caffeine can in fact potentially assist in preventing CBD effects on memory.

Another little bit of research, that one posted in the Addictive Behaviors journal, discovered that CBD "significantly reduced" (just as much as 40 percent) the number of cigarettes smoked by people who wished to quit.

CBD and Drowsiness

Another potential threat of CBD is it can cause drowsiness. Therefore, if you're taking supplements or medications to help you sleep, CBD could enhance these effects even more. It's important to keep this at heart if you're performing activities that want maximum alertness, such as driving or operating heavy machinery.

CBD Dosing Considerations

If you'd prefer to use CBD essential oil to help ease your arthritis pain, you might be wondering how much to manage or apply. However, when discussing proper CBD dosing, it's important to first understand bioavailability.

CBD Bioavailability

Bioavailability is thought as the "amount of the substance that becomes available (reaches the prospective organ or system circulation) for an organism's body for bioactivity when introduced through ingestion, inhalation, injection, or skin contact." In a nutshell, it is how a lot of the substance the body can effectively use predicated on the sort and route taken.

Regarding CBD oil specifically, Medscape shares that while intranasal administration of CBD has a bioavailability of 34 to 46 percent and CBD vaporization has 40 percent bioavailability, "oral CBD is regarded as only 6 percent, due to significant first pass-metabolism."

First-Pass Metabolism and CBD

Articles published in the journal Nurse Prescribing explains that first-pass metabolism occurs in the gut or the liver and is when a few of the drug is destroyed before with the ability to be circulated around your body. Thus, predicated on the research above, taking CBD oil in pill form may mean that your body only receives as little as 6 percent of the CBD.

Another option is to consider CBD essential oil topically, either only or within a lotion or cream. Research published in the European Journal of Pain studied the potency of this process and discovered that "topical CBD application has therapeutic prospect of relief of arthritis pain-related behaviors and inflammation without evident side-effects."

Proper CBD Dosing

Because CBD is so new, Medscape recommends that users "start low, go slow." This allows you to observe how well the body responds without giving it more than the required amount.

Plus, there are no government regulations about the manufacture and sale of CBD, and that means you never know if you're getting the total amount actually listed on the label. This makes erring privately of caution even more important.

CBD Essential oil Options

Consumer Health Break down (CHD) researched products on the internet and, according to the guide, the 3 most significant terms to learn before purchasing your CBD oil are CBD volume, hemp oil volume, and CBD concentration. The next and third will be the most significant, according to CHD, because they indicate the oil's potency.

It's also important to understand that we now have different product options in regards to CBD natural oils. They may be:

- CBD essential oil, which is the "stronger version and delivery system of CBD"

- CBD essential oil topical solutions, allowing an individual to utilize it on a far more targeted area.

- CBD essential oil tinctures, which are alcohol-based cannabis extracts that may be ingested orally, and typically the most popular CBD essential oil choice.

- CBD health supplements, or tablets containing CBD essential oil in a powdered form

The one you select will depend mainly on the place and severity of your pain. If you would like to target a particular area, an essential oil or essential oil topical could be the best. However, if the pain is all over, then the tincture or health supplement may supply the most relief.

Also, if you're not used to using CBD oil, we recommend you focus on 10 milligrams each day to determine a safe, effective amount for you. The majority of our RA colleagues finish up getting relief when they take 20mg of full spectrum lab grade CBD tincture under the tongue two times per day. It's important to carry it there for at least 60 seconds. Others with an increase of severe pain go up to 80mg twice or even more each day to get full relief.

Where you can Buy Lab Grade CBD Oil and Topical CBD Cream?

There's been an explosion of new CBD products hitting the marketplace before year. We're viewing reports of several of these having much less CBD than they stated or even including no CBD whatsoever. Worse, some have even failed testing for pesticides and dangerous bacteria. The FDA appears to be ramping up enforcement to tackle this issue. We tested a large number of products among our staff, and through our network of RA readers, found Spruce CBD to be the very best.

chapter 3

Benefits

Folks who have used CBD traditionally used it for a large number of years to take care of numerous kinds of pain, however the medical community have only recently begun to review it again.

Below are a few of the possible advantages of CBD oil:

- Arthritis pain

Seniors man's hand, one hand keeping the thumb of the other credited to arthritis pain.

CBD essential oil is popular for easing pain associated with arthritis.

A report in the Western Journal of Pain used an animal model to find out if CBD may help people who have arthritis manage their pain. Experts applied a topical gel comprising CBD to rats with arthritis for 4 times.

Their researchers note a substantial drop inflammation

and signals of pain, without additional side effects.

People using CBD essential oil for arthritis could find rest from their pain, but more human being studies have to be done to verify these findings.

Multiple sclerosis

Multiple sclerosis (MS) can be an autoimmune disease that impacts the whole body through the nerves and brain.

Muscle spasms are one of the most typical symptoms of MS. These spasms can be so excellent that they cause continuous pain in a few people.

One report discovered that short-term use of CBD essential oil could decrease the degrees of spasticity a person feels. The email address details are moderate, but many people reported a decrease in symptoms. More studies on humans are needed to confirm these results.

- Chronic pain

The same report studied CBD use for general chronic pain. Analysts compiled the results of multiple organized reviews covering a large number of studies. Their research figured there is considerable proof that cannabis is a highly effective treatment for chronic pain in adults.

A separate research in the Journal of Experimental Medication helps these results. This research shows that using CBD can decrease pain and swelling.

The experts also discovered that subjects weren't likely to build-up a tolerance to the consequences of CBD, so they might not want to increase their dosage continually.

They noted that cannabinoids, such as CBD, can offer helpful new treatments for individuals with chronic pain.

Other Uses

In America, CBD oil has varying legality across different states with a federal level, yet it currently has a variety of applications and encouraging possibilities.

Included in these are:

- smoking cessation and medication withdrawal.

- treating seizures and epilepsy.

- anxiety treatment.

- reducing a few of the consequences of Alzheimer's, as shown by initial research.

- antipsychotic effects on people who have schizophrenia.

- future applications in combating acne, type 1 diabetes, and cancer

Although more research must confirm some uses of CBD oil, it is shaping up as a possibly promising and versatile treatment.

In June 2018, the U.S. Food and Medication Administration (FDA) approved one form of CBD as cure for individuals with two uncommon and specific types of epilepsy, namely Lennox-Gastaut symptoms (LGS) or Dravet symptoms (DS). The brand of the drug is Epidiolex.

CHAPTER 5

Dosage

The FDA will not regulate CBD for some conditions. Because of this, dosages are available to interpretation, and folks should treat them with extreme caution.

Anyone who desires to use CBD should first talk with a health care provider about whether it's a good notion, and exactly how much to consider.

The FDA recently approved a purified form of CBD for a few types of epilepsy, with the brand Epidiolex. If you work with this medication, make sure to check out the doctor's advice about dosages.

Side effects

Exhausted businessman at desk rubbing eye, credited to headache and fatigue.

Possible short-term side ramifications of using CBD oil include fatigue and changes in appetite.

A lot of people tolerate CBD essential oil well, but there are a few possible side results.

According to an assessment in Cannabis and Cannabinoid Research, the most typical side results include:

- tiredness.

- Diarrhea.

- changes in appetite.

- putting on weight or weight loss

Furthermore, using CBD oil with other medications could make those medications pretty much effective.

The review also notes that scientists have yet to review some areas of CBD, such as its long-term effects on hormones. Further long-term studies will be helpful in identifying any side results CBD is wearing the body as time passes.

Folks who are considering using CBD essential oil should discuss this using their doctors. Doctors would want to monitor the individual for just about any changes and make modifications accordingly.

The individual information leaflet for Epidiolex cautions that there surely is a threat of liver harm, lethargy, and perhaps depression and thoughts of suicide, but they are true of other treatments for epilepsy, too.

CBD and other cannabinoids could also put an individual in danger for lung problems.

One research in Frontiers in Pharmacology, suggested cannabinoids' anti-inflammatory impact may reduce irritation too much.

chapter 4

CBD Essential oil for Relief

CBD essential oil works to alleviate discomfort in amazing ways. Actually, scientists are uncovering that the endocannabinoid system interacts at many factors with this major physical control systems. This natural system is mixed up in support of overall comfort, and cannabinoids that react to receptors in the machine have the ability to work to ease discomfort.

Many reports have analyzed the results of cannabis chemical substances, including both THC and CBD. We've learned out of this research that CBD essential oil can alleviate various kinds of pain by getting together with receptors inside our brains.

One part of comfort often associated with CBD oil is periodic joint discomfort. While more research must be conducted, some research demonstrates it might be a useful method of assisting overall joint comfort.

How exactly to Use CBD Essential oil for Pain - Dose Information

There are many techniques you may use CBD oil for pain. Here's a breakdown of what's available and can be utilized orally or topically for treatment:

- Oils: The very best CBD natural oils are full spectrum, meaning they include all substances found naturally in the place, like the cannabinoids (with track levels of THC), terpenes and essential natural oils. You'll find CBD natural oils in a container with a dropper. This enables you to ingest the essential oil by putting it under your tongue, allowing it to sit for approximately 15 mere seconds and then swallowing it.

- Tinctures: Tinctures are another popular way to use CBD, likely because you can simply gauge how much cannabidiol you are ingesting, like CBD essential oil. A tincture is usually extracted with alcoholic beverages or another solvent. Having a tincture, you utilize a dropper and place the drops under your tongue. Sometimes, manufacturers use carrier natural oils, natural flavors or fatty natural oils in their tinctures.

- Pills: CBD pills can be studied orally with drinking water. You'll have the ability to find CBD tablets in a variety of dosages, typically in levels of 10-40 milligrams. Tablets can be studied once to 3x daily, depending on your intensity of pain.

- Powder: CBD powder can be put into smoothies, juices, drinking water or any kind of drink. Check the label for dose instructions.

- Topical solution: Topical salves, lotions and gels containing CBD can be found and can be employed directly to regions of pain, like the back, neck, knees, hands and feet. Browse the product label for directions and strength.

Just how much CBD essential oil in the event can you use for pain? There is absolutely no official meal for CBD essential oil because everyone reacts in a different way to cannabinoids. To be able to determine the right

CBD essential oil dosage to ease your pain, you'll need to have a few things into consideration.

What are your targets? To be able to set up the right CBD essential oil medication dosage for you, you'll need to pinpoint your wellbeing goals. If you're looking for treatment, specify which kind of pain you are wishing to lessen. This will help you to monitor the effectiveness of your preliminary CBD dosage and determine if it's performing.

What's your severity of pain? Your CBD dose depends on your situation, which means you (together with your doctor) should categorize your individual needs as low, media, high or high.

Low - focus on a 5-10 milligram dose

Medium - focus on a 10-20 milligram dose

High or high - focus on a 20-40 milligram dose

Is my starting medication dosage working? Once you as well as your doctor have pinpointed your unique goals and decided your position, you'll focus on a short CBD dosage.

It's better to begin with lower dosages and work the right path up. Why? Because everyone has a different level of sensitivity to cannabis substances. Some individuals only need an extremely small amount to note the beneficial results, while others might need higher doses.

Start on the low end of your recommended dose range and take that amount regularly for 3-7 times. Keep going back again to your preliminary goals and assess whether this starting dosage is enhancing your symptoms.

It's working? Great! Stick to this dose, which may be used 1-3 times daily or relating to directions as well as your healthcare professional.

Don't notice any improvements yet? Then consult with your doctor to see when you can increase your dosage by 5 milligrams and stick to this new amount for another 3-7 times. Keep this routine heading until you've found the right CBD medication dosage for you.

How long will it take CBD oil to work for treatment? It's

generally suggested that you utilize CBD essential oil about 1 hour before desired benefits. Typically, you'll spot the results within 30-60 minutes.

The advantages of CBD oil for pain can last about 4-6 hours, with respect to the dosage. When you have other needs, you might take advantage of taking a dosage every six hours roughly.

CHAPTER 6

CBD Oil Unwanted Effects and Precautions

The existing research shows that CBD utilization has few and generally slight side effects, and a "tolerance" for CBD will not seem that occurs. If you are taking a look at CBD vs. THC, it's THC that gets the mind-altering results that makes you feeling "high." CBD doesn't have intoxicating results.

Right now, there are a great number of CBD products on the marketplace, and that means you need to look carefully. Only use CBD essential oil products that are examined for contaminants and show CBD vs. THC levels. Choose a product that has received a COA, or certificate of evaluation, which means that it's been examined and met lab requirements.

It's also strongly suggested that you decide to go with a natural CBD essential oil. The hemp herb is a "bioaccumulator," meaning it's with the capacity of absorbing toxins from the ground, drinking water and air faster than the pace they're lost. So, heading organic for all your CBD products will make sure that you aren't also ingesting harmful pesticides and other chemicals.

7 Benefits and Uses of CBD Essential oil (Plus Unwanted Effects)

Healthline and our companions may get a portion of income if you make a purchase utilizing a link upon this page.

Cannabidiol is a favorite natural treatment used for most common ailments.

Better known as CBD, it is one of the 104 chemical substances known as cannabinoids within the cannabis or weed herb, Cannabis sativa (1Trusted Source).

Tetrahydrocannabinol (THC) is the primary psychoactive cannabinoid within cannabis, and causes the feeling to getting "high" that's often associated with cannabis. However, unlike THC, CBD is not psychoactive.

This quality makes CBD an attractive option for individuals who want for rest from pain and other symptoms with no mind-altering ramifications of marijuana or certain pharmaceutical drugs.

CBD oil is manufactured by extracting CBD from the cannabis flower, then diluting it with a carrier essential oil like coconut or hemp seed essential oil.

It's gaining momentum in medical and wellness world, with some scientific tests confirming it could help treat a number of illnesses like chronic pain and stress and anxiety.

Listed below are seven health advantages of CBD oil that are backed by scientific evidence.

1. Can Decrease Pain

Cannabis has been used to take care of pain dating back

to 2900 B.C. (2Trusted Source).

More recently, researchers have discovered that particular components of weed, including CBD, are accountable for its pain-relieving results.

The body contains a specific system called the endocannabinoid system (ECS), which is involved with regulating a number of functions including sleep, appetite, pain and disease fighting capability response (3Trusted Source).

Your body produces endocannabinoids, that are neurotransmitters that bind to cannabinoid receptors in your anxious system.

Studies show that CBD can help reduce chronic pain by impacting endocannabinoid receptor activity, lowering inflammation and getting together with neurotransmitters (4Trusted Source).

For instance, one research in rats discovered that CBD

injections reduced pain response to surgical incision, while another rat research found that dental CBD treatment significantly reduced sciatic nerve pain and inflammation (5Trusted Source, 6Trusted Source).

Several individual studies have discovered that a mixture of CBD and THC works well in treating pain related to multiple sclerosis and arthritis.

An oral aerosol called Sativex, which really is a mixture of THC and CBD, is approved in a number of countries to take care of pain related to multiple sclerosis.

In a report of 47 people who have multiple sclerosis, those treated with Sativex for just one month experienced a substantial improvement in pain, walking and muscle spasms, set alongside the placebo group (7Trusted Source).

Another study discovered that Sativex significantly improved pain during motion, pain at rest and rest quality in 58 people who have arthritis rheumatoid (8Trusted

Source).

CBD, especially in mixture with THC, may succeed in lowering pain associated with diseases like multiple sclerosis and arthritis rheumatoid.

2. Could Reduce Stress and Depression

Panic and depression are normal mental health disorders that can have devastating influences on health insurance and well-being.

Based on the World Health Business, depression is the sole largest contributor to impairment worldwide, while nervousness disorders are ranked sixth (9Trusted Source).

Anxiousness and depression are usually treated with pharmaceutical drugs, which can result in a number of aspect results including drowsiness, agitation, insomnia, sexual dysfunction and headaches (10Trusted Source).

What's more, medications like benzodiazepines can be addictive and could lead to drug abuse (11Trusted Source).

CBD oil shows promise as cure for both depression and stress, leading many who live with these disorders to be thinking about this natural strategy.

In one research, 24 people who have social panic received either 600 mg of CBD or a placebo before a presenting and public speaking test.

The group that received the CBD had considerably less anxiety, cognitive impairment and distress in their speech performance, set alongside the placebo group (12Trusted Source).

CBD essential oil has even been used to safely treat insomnia and panic in children with post-traumatic stress disorder (13Trusted Source).

CBD in addition has shown antidepressant-like results in several pet studies (14Trusted Source, 15Trusted Source).

These characteristics are associated with CBD's ability to do something on the brain's receptors for serotonin, a neurotransmitter that regulates disposition and interpersonal behavior.

Using CBD has been proven to reduce anxiousness and depression in both human being and pet studies.

3. Can Alleviate Cancer-Related Symptoms

CBD can help reduce symptoms related to tumor and side results related to cancers treatment, like nausea, vomiting and pain.

One study viewed the consequences of CBD and THC in 177 people who have cancer-related pain who didn't experience rest from pain medication.

Those treated with an extract containing both materials experienced a substantial decrease in pain in comparison to those who received only THC extract (16Trusted Source).

CBD also may help reduce chemotherapy-induced nausea and vomiting, that are among the most typical chemotherapy-related side results for people that have malignancy (17Trusted Source).

Though there are drugs that assist with these distressing symptoms, they are occasionally ineffective, leading many people to get alternatives.

A report of 16 people undergoing chemotherapy discovered that an one-to-one mixture of CBD and THC administered via mouth area squirt reduced chemotherapy-related nausea and vomiting much better than standard treatment alone (18Trusted Source).

Some test-tube and animal studies have even shown that CBD may have anticancer properties. For instance, one test-tube research found that focused CBD induced cell loss of life in human breasts tumor cells (19Trusted Source).

Another research showed that CBD inhibited the pass on of aggressive breasts cancers cells in mice (20Trusted Source).

However, they are test-tube and animal studies, to allow them to only suggest what may work in people. More studies in humans are needed before conclusions can be produced.

Though CBD has been proven in reducing symptoms related to cancer and cancer treatment, and could have even cancer-fighting properties, more research is required to evaluate its efficacy and safety.

4. May Reduce Acne

Acne is a common condition of the skin that impacts more than 9% of the populace (21Trusted Source).

It is regarded as the effect of a quantity of factors, including genetics, bacteria, underlying swelling and the overproduction of sebum, an oily secretion created by sebaceous glands in your skin (22Trusted Source, 23).

Predicated on recent scientific tests, CBD oil can help treat acne because of its anti-inflammatory properties and ability to lessen sebum production.

One test-tube research discovered that CBD essential oil prevented sebaceous gland cells from secreting excessive sebum, exerted anti-inflammatory activities and prevented the activation of "pro-acne" brokers like inflammatory cytokines (24Trusted Source).

Another research had similar findings, concluding that CBD may be a competent and safe way to take care of acne, thanks partly to its impressive anti-inflammatory characteristics (25Trusted Source).

Though these email address details are encouraging, human studies exploring the consequences of CBD on acne are needed.

CBD may have beneficial results on acne because of its anti-inflammatory characteristics and its capability to regulate the overproduction of sebum from the sebaceous glands.

5. May Have Neuroprotective Properties

Researchers think that CBD's capability to do something on the endocannabinoid system and other brain signaling systems might provide benefits for people that have neurological disorders.

Actually, one of the very most analyzed uses for CBD is within treating neurological disorders like epilepsy and multiple sclerosis. Though research in this field continues to be relatively new, several studies show promising results.

Sativex, a dental spray comprising CBD and THC, has shown to be always an effective and safe way to lessen muscle spasticity in people who have multiple sclerosis.

One study discovered that Sativex reduced spasms in 75% of 276 people who have multiple sclerosis who have been experiencing muscle spasticity that was resistant to medications (26Trusted Source).

Another research gave 214 people who have severe epilepsy 0.9-2.3 grams of CBD oil per pound (2-5 g/kg) of bodyweight. Their seizures reduced with a median of 36.5% (27Trusted Source).

One more research discovered that CBD essential oil significantly reduced seizure activity in children with Dravet symptoms, a complex child years epilepsy disorder, in comparison to a placebo (28Trusted Source).

However, it's important to notice that some individuals in both these studies experienced effects associated with CBD treatment, such as convulsions, fever and diarrhea.

CBD in addition has been researched because of its potential performance in treating other neurological diseases.

For instance, several studies show that treatment with CBD improved standard of living and rest quality for individuals with Parkinson's disease (29Trusted Source, 30Trusted Source).

Additionally, animal and test-tube studies show that CBD may decrease inflammation and assist in preventing the neurodegeneration associated with Alzheimer's disease (31Trusted Source).

In a single long-term research, researchers gave CBD to mice genetically predisposed to Alzheimer's disease, discovering that it helped prevent cognitive decrease (32Trusted Source).

Though research is bound at the moment, CBD has been proven to effectively treat symptoms related to epilepsy and Parkinson's disease. CBD was also proven to reduce the development of Alzheimer's disease in test-tube and pet studies.

6. Could Benefit Center Health

Recent research has connected CBD with many perks for the heart and circulatory system, like the ability to lessen high blood circulation pressure.

High blood circulation pressure is associated with higher risks of lots of health issues, including stroke, coronary attack and metabolic symptoms (33Trusted Source).

Studies indicate that CBD may be considered an effective and natural treatment for high blood circulation pressure.

One recent research treated 10 healthy men with one dosage of 600 mg of CBD essential oil and found it reduced resting blood circulation pressure, in comparison to a placebo.

The same study also gave the men stress tests that

normally increase blood circulation pressure. Interestingly, the solitary dosage of CBD led the men to see a smaller blood circulation pressure increase than normal in response to these assessments (34Trusted Source).

Research workers have suggested that the stress and anxiety-reducing properties of CBD are accountable for its capability to help lower blood circulation pressure.

Additionally, several animal studies have demonstrated that CBD can help decrease the inflammation and cell death associated with cardiovascular disease because of its powerful antioxidant and stress-reducing properties.

For instance, one study discovered that treatment with CBD reduced oxidative stress and prevented center harm in diabetic mice with cardiovascular disease (35Trusted Source).

Though more human studies are needed, CBD may benefit heart health in a number of ways, including by reducing blood circulation pressure and avoiding heart damage.

Healthline Partner Solutions

Get Answers from a health care provider in Minutes, anytime.

Have medical questions? Connect to a board-certified, experienced doctor online or by telephone. Pediatricians and other specialists available 24/7.

7. OTHER Potential Benefits

CBD has been studied because of its role in treating lots of medical issues apart from those outlined above.

Though more studies are needed, CBD is considered to supply the following health advantages:

Antipsychotic effects: Studies claim that CBD can help people who have schizophrenia and other mental disorders by reducing psychotic symptoms (36Trusted Source).

Drug abuse treatment: CBD has been proven to change circuits in the mind related to medication habit. In rats, CBD has been proven to lessen morphine dependence and heroin-seeking behavior (37Trusted Source).

Anti-tumor results: In test-tube and pet studies, CBD has proven anti-tumor results. In animals, it's been shown to avoid the spread of breasts, prostate, brain, digestive tract and lung malignancy (38Trusted Source).

Diabetes avoidance: In diabetic mice, treatment with CBD reduced the occurrence of diabetes by 56% and significantly reduced inflammation

How exactly to Use CBD Essential oil for TREATMENT

Cannabis has been used to alleviate pain for a large number of years and CBD, one of the dynamic substances in the cannabis vegetable species, has that which can cause powerful analgesic and anti-inflammatory results.

In ancient Chinese language texts dating back again to 2900 B.C., cannabis is referred to as a kind of medication for rheumatic pain. The herb was also found in combination with wines to anesthetize patients during surgical treatments. And in India, around 1000 B.C., cannabis was appreciated as an analgesic, antispasmodic and anti-inflammatory agent.

Although some compounds of the cannabis plant were found in ancient medicine, like the psychoactive compound THC, we realize from newer studies that CBD is a pain-relieving powerhouse alone. Treatment is one of the very most well-known CBD essential oil benefits, and once and for all reason, as the substance works to suppress pain procedures and indicators in the mind.

If you need a natural, effective and safe method of relieving your pain, it's smart to consider using CBD essential oil. With chronic pain staying one of many causes of struggling and impairment in the world, it's time we turn to our ancestors for answers.

CBD Essential oil for Treatment: 7 Types of Pain that Are Helped

CBD essential oil works to alleviate pain by mediating pain neurons and initiating anti-inflammatory and analgesic results. Scientists are uncovering that the endocannabinoid system interacts at many factors with this major pain control systems. This natural system is mixed up in control of pain, and cannabinoids that react to receptors in the machine have the ability to modulate pain awareness.

Many reports have analyzed the results of cannabis chemical substances on treatment, including both THC and CBD. We've learned out of this research that CBD essential oil can alleviate various kinds of pain by getting together with receptors inside our brains and reducing irritation.

This is a rundown of the types of pain that may be managed with CBD oil:

- Joint Pain

CBD essential oil works to alleviate joint swelling and pain. Research implies that it could be a useful restorative agent for dealing with joint neuropathic pain. Research conducted at Northwestern University College in Chicago signifies that cannabis for joint pain is the biggest medically obtained utilization of the seed.

Although controlled clinical trials involving humans had a need to prove the efficacy of CBD for rheumatic diseases, researchers remember that preclinical and individual data that do exist indicate that the utilization of cannabis materials, including CBD, should be studied seriously as a potential treatment for joint pain.

Within an animal study published in Pain, local administration of CBD on rats blocked osteoarthritis pain and avoided the later development of pain and nerve damage in osteoarthritic joints.

- Nerve Pain

Nerve pain, also called neuropathic pain, is triggered by harm to the body's nervous system. In pet studies, cannabinoids could actually attenuate neuropathic pain that was made by traumatic nerve damage, disease and poisonous insults.

And when the consequences of medicinal cannabis arrangements containing both THC and CBD were applied to humans with neuropathic pain, clinical studies

63

affirm the positive advantages of cannabinoids for nerve treatment.

- Back Pain

CBD essential oil produces pain-inhibitory results by getting together with receptors in the human brain that are accountable for the pain response. Revitalizing receptors in the endocannabinoid system reduces the strength of back again pain, and other kinds of pain, and supports the reduced amount of inflammation.

Research shows that cannabinoids work for the treating of chronic pain conditions, like back again pain, and in comparison to placebo, are associated with a decrease in patient pain rankings.

- Knee Pain

Exactly like CBD's pain-relieving effects on back again pain, additionally, it may help alleviate knee pain by getting together with receptors that initiate action inside our pain nerve fibers. This is one way CBD causes analgesic results and can help people experiencing various kinds of pain.

CBD oil also offers anti-inflammatory effects that will help to reduce leg swelling, tenderness and stiffness, according to a report conducted at the College or University of South Carolina's College of Medicine.

Acknowledgments

The Glory of this book success goes to God Almighty and my beautiful Family, Fans, Readers & well-wishers, Customers and Friends for their endless support and encouragements.

Printed in the USA
CPSIA information can be obtained
at www.ICGtesting.com
LVHW040354220124
769411LV00104B/1108

9 781685 220129